# Oil People

## Natalie Bright

Apollo Publishing L.L.C.
Amarillo, Texas

Dedicated to
The hard-working, committed people
of the oil industry.

© 2010 Natalie Bright
All rights reserved. Published 2010

Apollo Publishing L.L.C.
1415 23rd Street
Canyon, Texas 79015

www.oilpeople.net

Photography: Joe Don Stevens
Editor: Melanie Mallon  www.malloneditorial.com
Layout: Westcom Associates  www.Help-U-Publish.com

18 17 16 15 14 13 12 11 10   1 2 3 4 5

ISBN: 978-0-615-35949-6 (Library Binding)

Printed in China

All rights reserved. No part of this publication may be reproduced, stored in a retrieval system, or transmitted, in any form or by any means, electronic, mechanical, photocopying, recording, or otherwise without the prior written permission of the author.

## Table of Contents

1. The Crew — 4
2. The Job Site — 8
3. The 42-Gallon Barrel — 10
4. What's in a Barrel? — 11
5. The Early Ages — 12
6. Protecting our Environment — 18
7. Safety First — 19
8. A Global Product — 20
9. The Search Continues — 23
10. What about *My* Future? — 26

Glossary — 29

Recommended Websites for More Information — 31

Acknowledgements — 31

Selected Bibliography — 31

# 1            The Crew

The stakes are high. The work can be deadly. If they fail, the loss is enormous. What they seek is hidden miles below the earth's surface. They are men and women with a common goal: discover, refine, and transport crude oil.

Oil people represent a varied workforce of skilled laborers and scientists who are all needed to bring in and complete just one successful oil well.

> Millions of mothers and fathers, aunts and uncles, grandfathers and grandmothers work in the oil industry!

A **petroleum geologist** is the scientist who proposes the spot to drill. The earth is his domain. By applying what he knows about rock formations and the earth's processes, he can pinpoint the best possible location to find crude oil. This study can take several years before a well is actually drilled.

Once the best drilling site is determined, a landman is sent to lease the minerals in the land. A lease is signed permission from the landowner allowing a well to be drilled on the owner's property. The landman works out an agreement for payment. The landowner receives money for surface damages and for a percentage of the sales of all oil taken

from under his land's surface. This is called a royalty.

Before a drilling rig is moved to the location, a surveyor marks the exact spot for drilling based on the geologist's directions. After the surveyor has set his flag, the earth movers

use their bulldozers to build the location. The drilling site must be leveled to specific calculations.

The drilling crew brings the drilling rig and necessary equipment on trucks. Crews usually consist of four men for every twelve-hour shift. Their boss is called a tool pusher.

**Bakersfield, California**
photo by Christina Rothman

Steel **casing pipe** as well as joints of **drill pipe** are delivered by the supplier and stacked on pipe racks.

A welder is needed to repair anything that may have broken during transport and to weld the blow out preventer in place. This device is joined to the **casing pipe** to prevent high-pressured fluid from rushing to the surface.

The order is given to **spud** and drilling begins. A drilling **engineer** establishes a procedure for this part of the operation.

Wildcatter is a person who proposes or funds a well in a location not previously explored for oil.

The mud man delivers sacks of drilling mud and carefully monitors the thickness of the mud during drilling. This mud is not just any kind of dirt. Drilling mud is a special soil mined in Wyoming and is composed of bentonite clay.

The company man oversees the entire operation

**A drilling crew makes a pipe connection.**

from start to finish. He is in charge of every worker at the drilling site. It is his job to make certain that the operation runs smoothly and that all safety procedures are followed.

> The thickness of the drilling mud is increased with cotton seed hulls, cedar fibers, or recycled shredded paper.

After the drilling has reached TD (total depth), a completion engineer directs the well's completion process. It is his job to ensure that the crude oil is successfully brought to the surface.

Roustabout crews are very useful workers who are knowledgeable about how to repair every piece of equipment on location. They design and bury a flowline system, which carries the **crude** to storage tanks called a **tank battery**. The oil stays here until a transport trucker picks it up and hauls the load to the **refinery**.

The operator is the company that directs all field activity and pays the workers. The operator hires a pumper, who checks on the well every day to make certain oil is flowing to the surface.

> Worms are new employees to oil field work.

A pipeline, or a transport trucker, hauls the oil from the tank battery to the refinery, where it is made into the products that we use.

# 2                          The Job Site

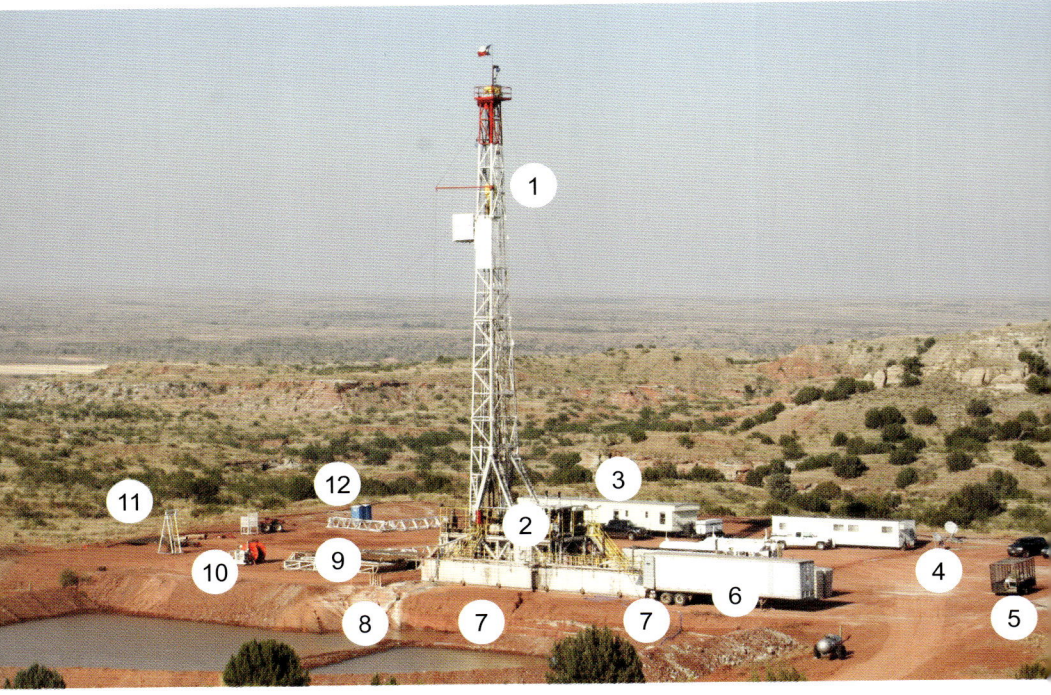

**Key to Drilling Location**

| | | |
|---|---|---|
| 1. Derrick | 5. Trash Trailer | 9. Pipe Racks |
| 2. Doghouse | 6. Cotton Seed Hulls Van | 10. Drilling Line Spo[ol] |
| 3. Company Man's Trailer | 7. Drilling Mud Pits | 11. Headache Rack |
| 4. Communication Satellite | 8. Reserve Pits | 12. Portable Potty |

Some oil people call a bright blue sky and a patch of scrub brush their office.

Even when the clouds turn dark and ugly, bringing a promise of rain or snow, the drill bit keeps turning. To stop might risk damage to the fragile rock formations deep below the earth's surface.

Some oil workers spend hours reading land records in courthouse basements. Others work in laboratories surrounded by core samples and high-tech computers. Many labor in steel cities of pipelines and storage tanks, refining crude into the products the world uses.

---

Oil wells can range in depth from several thousand feet to several miles deep. Depending on the rock formations and conditions, drilling rigs generally drill about 1,000 feet per day.

---

Crews cook and eat at the rig.
This is one consultant's special recipe.

### SEQUOYAH'S BEANS

| | |
|---|---|
| 2 large cans ranch beans | 6 large cloves garlic, chopped |
| 2 large cans pinto beans | 1 large onion, chopped |
| 1 can tomatoes & green chilies | 1 tablespoon garlic powder |
| 1/3 cup chopped fresh cilantro | 1 tablespoon onion powder |
| | 5 shakes Worcestershire |

Simmer all ingredients in a big pot on low heat, stirring often.

# 3          The 42-Gallon Barrel

Oil is measured in barrels because that was the only way to store it and transport it in the early days. A 42-gallon wooden barrel was a standard size used for many things, such as pickles, syrup, and butter.

During the early days of the industry, barrels of every shape and size were used to hold the constant flow of **petroleum**. Barrel making became an important profession.

Using teams of up to twenty horses, teamsters hauled crude on wagons along rutted roads and over rocky hills to rivers or to railheads. River men or railroaders took over from there, delivering crude oil to the refineries for processing.

Today, the transporting of **crude** has grown into a complex system of buried pipelines, railcars, trucks, and ships.

**Oil barges and barrels on the Alleghany River**
from Pennsylvania Historical and Museum Commission
Drake Well Museum Collection, Titusville, PA

# 4        What's in a Barrel?

## Product and Gallons per Barrel

Gasoline 19.4
Heating Oil and Diesel Fuel 10.5
Kerosene-Type Jet Fuel 4.1
Coke (carbon) 2.2
Heavy Fuel Oils 1.7
Liquefied **Refinery** Gases 1.5
Still Gas 1.8
Asphalt and Road Oil 1.4
Raw Material for Petrochemicals 1.1
Lubricants 0.4
Kerosene 0.2
Other 0.4

Figures from the American Petroleum Institute and based on average yields for U.S. refineries in 2005. One barrel contains 42 gallons of crude oil. The total volume of products made is 2.7 gallons greater than the original 42 gallons of crude oil. This represents *processing gain*.

---

### Today crude oil is not transported or stored in barrels. The 42-gallon barrel is a standard for measurement.

---

**Fiberglass tanks hold formation water and steel tanks hold crude oil at a tank battery in the Texas Panhandle.**

# 5 The Early Ages

Petroleum has been around for thousands of years. Once called bitumen, oil has flowed naturally to the earth's surface in many parts of the world.

Early man used the sticky goop to waterproof his reed baskets and canoes. Ancient Egyptians used bitumen to grease their chariot-wheel axles. Pitch--a black, thick crude--was mixed with gravel to pave Babylonian roads.

## Pennsylvania

In Pennsylvania the Seneca Indians treated their cuts and bruises with a dab of the foul-smelling crude. They soaked in the soothing waters of Oil Creek, so named because of the slick matter that floated on the surface. A torch to the water lighted their nighttime festivals.

This area of Pennsylvania was also known for its vast numbers of salt domes. A man from Tarentum decided to keep the crude oil that appeared in his father's salt wells and advertised it as a cure all medicine. Samuel Kier's Rock Oil sold for fifty cents a bottle.

Kier is also credited with initiating the process of distilling, or heating, the crude, which resulted in a pale gold oil. This new lamp oil burned with a brilliant light, and it was cheaper than the popular whale oil.

As talk of Sam Kier's success spread, investors began forming companies to look for crude. The most likely place to find more was where Native Americans had used the sticky stuff for hundreds of years.

Edwin Drake made his mark in history by drilling the first discovery well near Oil Creek 150 years ago. Known as a lively story-teller, Drake was easily recognized by his stove-pipe hat and long overcoat. While living at a bustling New Haven hotel, he formed a friendship with a local banker, James Townsend. Townsend asked Drake to consider the opportunities for starting an oil company in a place called Titusville.

**Drake Well**
from Pennsylvania Historical and Museum Commission, Drake Well Museum Collection, Titusville, PA

Drake settled his family in Titusville and hired a salt water driller from Tarentum. "Uncle" Billy Smith agreed to a wage of $25 per month, which included the labor of his 15-year old son, Samuel. Uncle Billy had drilling experience and, as a former blacksmith, could make the tools they would need. Instead of digging ditches and holes as was the custom, they decided to drill into the earth.

Despite teasing from the townspeople about their unusual methods, crude bubbled to the surface from 69 feet in 1859. Uncle Billy saw "a golden yellow oil, foaming with bubbles."

---
Brand New Model T Ford: $400
Tank of gasoline: $12.50
Driving to grandma's instead of walking:
PRICELESS
---

The search for crude moved across the country, as new uses for oil were realized. The vehicle powered by a

**Model T Ford**
Panhandle-Plains Historical Museum, Canyon, Texas (ca. 1915)

**A train load of crude heading to a refinery**
Panhandle-Plains Historical Museum, Canyon, Texas

combustion engine, invented by German engineer Gottlieb Daimler, shifted our ancestors from horsepower to fossil fuel power. Henry Ford's assembly line made it possible for more Americans to afford a gasoline-powered motor car.

Many significant crude discoveries were made in the United States, which proved that oil could be found in other places besides Pennsylvania and in underground formations other than salt domes.

The first shipload of unrefined oil traveled to Europe in 1861 with exports reaching 152 million gallons per year by 1872. United States crude journeyed across the ocean for other countries such as France, Syria, Egypt, and Great Britain. "The world is weeping for oil" was a popular phrase of the time.

---

Once found,
OIL is on the move
and never stops!

---

## Texas

A group of drilling contractors from Corsicana, Texas, brothers Curt, Allen, and James Hamill, drilled on a lease named Spindletop. The work was hard, and there were many problems along the way, but their hard labor paid off. Crude spewed out of the well, shooting over twice as high as the **derrick**. Oil flowed like rivers across the location. Men and mule teams struggled to build ditches and earthen tanks fast enough to contain the oil.

The Lucas No. 1 Well,
Spindletop Lease, 1901
Texas Energy Museum Beaumont, Texas

Crude spewed for nine days, drifting over the town of Beaumont and misting the houses with a dirty golden smear. The world had never seen a **gusher** before. The discovery well on the Spindletop lease came in at 100,000 barrels of oil per day, proving that oil could be found in huge quantities. The theories about how to search for oil had changed. Geologists began looking at geological structure.

## California

California became a significant producer of oil. One important find in the state was the Kern River Field. Over 24 million barrels came out of this area in 1903. The crude from Kern was dark and heavy, which made it perfect fuel for locomotives.

For a $10.60 round trip ticket, you could take a sight-seeing ride from San Francisco through the Kern River Oil Field on the Southern Pacific Railroad.

## Oklahoma

Ida Glenn, a Creek Indian, reported oil seeps on her land to wildcatter Robert Galbreath. Drilling began in September 1905, and Oklahoma's first major oil field gushed in from 1,400 feet. As part of the richest oil field the world had seen yet, the Ida E. Glenn Number One well proved that oil existed mid-continent.

**All dressed up for a sight-seeing tour of an oil field.**
Panhandle-Plains Historical Museum, Canyon, Texas

# 6    Protecting Our Environment

Awareness of our environment has changed since the first discovery wells 150 years ago. States have regulatory agencies to oversee all operations.

For example, drillers are required to seal off fresh groundwater zones by cementing steel casing into the drilled holes. This lines the hole with a protective barrier of steel.

---
**Can you name a place where crude oil seeps to the surface today?**
*(See answer page 32.)*

---

Measures are also taken to ensure the safety of animals in the area of drilling operations. Nets are placed over open pits to keep birds out. Fences are built around the drilling location and around the **pumpjack** to stop grazing cattle and deer from getting too close to dangerous equipment.

State agency field inspectors regularly visit drilling sites and productive leases. Office workers in district and state headquarters regulate the required paperwork.

The same drilling site from page 8 with a pump jack.
The dirt areas will be seeded with native grass in the spring.

# 7 Safety First

Every worker and visitor on a drilling site must wear a hard hat, safety goggles, and steel-toed boots. Workers on the drilling floor wear ear plugs to protect their ear drums from the loud noises of the rotating bit and rig machinery.

At the beginning of every shift, workers hold a safety meeting. Each time a new company arrives on location to start a different course of action, a safety meeting is held. All rules are reviewed.

## DANGER!

Oil field equipment can cause serious injury. **Pumping units** are set on timers that can start at any time. Oil and natural gas comes out of the ground under high pressure and is highly flammable, which means even a tiny spark can create a huge explosion. Do not light a match near oil field equipment. Do not climb on or play near the equipment. Stay out of fenced-in areas.

**A foreman checks pressure gauges inside the fenced-in area around a disposal well.**

# 8   A Global Product

Most everyone in the world is a consumer of oil or of one of its byproducts. Crude-based fuel keeps passengers and supplies moving in cars, trucks, trains, jet planes, and ships. Diesel-powered tractors, bulldozers, mowers, and backhoes tug, pull, carry, and build things that make our lives easier.

Even the things we use that are not byproducts of petroleum, such as goods made from wood or glass or iron, are produced or transported using some form of fossil-fueled energy.

> Around the world, airplanes account for 300,000 flights every 24 hours, using 67 million gallons of jet fuel.

**Refined gasoline and diesel are stored in tanks, waiting for transport by tanker trucks to the consumer.**

A worker sands the molded cylinders that will be used in photo processing machines.

At a plastics fabrication plant workers prepare plastic molds for the furnace.

---

Can you imagine life without some of these petroleum-based products?

### School
crayons, rulers, computers, glue, tape, plastic folders

### Home
food wraps, telephone, paint, trash cans

### Fun
football helmets, beach balls, sunglasses, flip flops

### Health
artificial limbs, pacemakers, soft contact lenses

### Pets
flea collars, dog leashes, bird feeders, Frisbees

Independents are the smaller producers who do not own refineries. They explore and produce crude, then sell their oil to the companies who own refineries.

Majors usually refers to companies in the United States that own refineries and control a very small portion of the world's **reserves** (less than 10%).

We use every drop of oil that we produce in this country. The remainder of our oil is imported, or brought in to the U.S., to meet the demands of modern-day life.

The largest supplier of oil to our country today is Canada.

The majority of oil reserves in the world today is controlled by companies owned by foreign governments. A group called OPEC, the Organization of the Petroleum Exporting Countries, includes countries in the Middle East as well as Venezuela.

Top Ten
Oil Producing States
#1 Alaska
#2 Texas
#3 California
#4 Louisiana
#5 New Mexico
#6 Oklahoma
#7 Wyoming
#8 Kansas
#9 North Dakota
#10 Montana

**A crude oil refinery in the Texas Panhandle lights up the night sky.**

# 9 The Search Continues

Many years ago, ancient seas teemed with life such as the jawless fish, simple mollusks, primitive sponges, algae, plankton, and millions of trilobites. As these animals died and shed their outer shells, the bottoms of the seas became littered with their remains. On top of that, countless numbers of streams and rivers emptied sediment into the oceans. As animals and sediment piled up on sea floors, the heat and the pressure squeezed the decaying matter and generated fossil fuels.

**The hunt for crude oil is propelled by science and technology from a wide variety of professional skills.**

The petroleum migrated, or moved, into cracks, crevices, and layers of sand until it became trapped. In other cases, the source rock became the trap. Oil is found in sedimentary layers made up of shale, sandstone, or limestone. These formations were once riverbeds, seashores, or reefs, which are now buried beneath layers of sediment.

The Gulf of Mexico is a good example of this process. The gulf was once an inland sea where rivers from the surrounding land masses carried silt, sediment, and sand. Over millions of years, the build up of dying sea creatures and silt caused the floor to sink and become buried under layers. This is one of the most productive offshore regions in our country.

A specialized group of geologists, geophysicists, and engineers have devoted their life's work to the thrill of the hunt and the gamble of drilling a successful well.

Most exploration geologists believe we may have untapped reserves in our country waiting to be discovered. The only way to know for certain is to drill.

Combining facts about the earth's processes with data from all available resources, geologists and engineers follow the elusive trail of petroleum. They apply their years of experience and knowledge about rocks, physics, chemistry, other sciences, and mathematics to successfully produce reservoirs of fossil fuels.

Each **reservoir**, or trap, usually contains salt water, oil, and natural gas. If a well has too much salt water it might be considered a dry hole and uneconomical to produce. The amounts of salt water, oil, and natural gas are never the same proportion in any given formation.

Based on the knowledge of the earth's processes, the exploration geologist starts on the surface by studying rock formations present on our planet to consider what might lie beneath. Armed with various instruments, all available data is used to determine the most likely place for drilling a successful, productive oil well.

**A derrick man makes a pipe connection ten stories above the ground.**

# Searching for crude oil:

- The petroleum geologist gathers data such as logs and **scout tickets** from previously drilled wells, which describes any signs of oil and gas.
- All raw data are put into an orderly, logical format, usually in map form. This illustrates the general potential of areas where petroleum might be accumulated.
- Knowledge of geological processes are used to refine the map to include locations of specific geoforms, or rock formations, known to accumulate petroleum (reservoirs).
- The geologist must be confident that all necessary components of an oilfield are present: petroleum source rock; reservoir rock, such as sand or a reef that can contain the oil; and a **trapping mechanism**.
- If all components appear to be present, the geologist may request that a geophysicist run **seismic lines** to confirm the presence of structure and reservoir.
- Other tools may include geochemistry, magnetics, gravity surveys, and aerial and satellite imagery.
- None of these tools give absolute information, and all data are subject to interpretive error, meaning it can be misread. The only reliable oil-finding tool is the drill bit, but drilling is expensive, and the risk is high. The reputation of the petroleum geologist is key to proceeding to the drilling phase.

We do not have cameras that can provide a picture of what might be below the earth's surface. The only way to know for certain is to drill.

# 10  What about My Future?

Energy gets you to where you want to go. Energy keeps you warm and cools you off. Energy powers the movie screen and cooks the popcorn.

Energy from fossil fuels are non-renewable, which means the earth is not making any more.

Renewable energy can be used and made again.

Where will we find the energy to meet the demands of a growing world population?

The answer is everywhere.

Fossil fuels will continue to play an important role. The systems to transport and refine crude oil have been in place for 150 years. Other sources of energy will be important for our future too--maybe even some that have not been invented yet.

Dedicated scientists continue to search for efficient and effective methods of renewable energy to support our ever-changing modern world. Every alternative offers new challenges.

Solar Energy, or power from the sun, is one way to heat your home or cook your food. Challenge: Costly solar panels and collectors, and periods of limited sunshine due to nighttime, rain clouds, or pollution.

---
It takes energy to make more energy!
---

Hydropower, or energy produced by water, is used in some states to generate electricity. About 19% of electricity is produced by hydropower. Challenge: Ecological areas that will support this type of power generation without damaging fish habitats and displacing populated areas can be difficult to find.

Biogas is a byproduct of the estimated 1,700 active landfills in our country. As our garbage decomposes, it creates methane gas, which can be collected and reused to generate electricity or steam. Challenge: This technology requires available land for new landfills, and transporting trash from urban areas can be costly.

Ethanol is a biofuel made from starch crops, like corn, barley, and wheat. Challenge: Corn is raised on farms across the country, but it needs huge amounts of water to grow. The increased demand for corn increases the farmers' demand for diesel, a byproduct of petroleum. Diesel powers the tractors and trucks used for planting, harvesting, and transporting the corn to mills.

**A corn harvester looks like a giant spider as it fills up a waiting hauler.**

Wind turbines are used to generate electricity. Wind energy is fast becoming popular all over the world. Challenge: The wind does not blow every day, making this source of energy sometimes unreliable.

Geologists, engineers, and researchers devote their life's work to exploring for oil as well as solving the challenges of discovering alternative sources of energy. The future will include all sources of energy, as well as more efficient ways of using fossil fuels to meet our needs.

**One blade on a wind turbine tower is 131 feet long. Can you find the boy in the picture? Hint: look for a blue shirt.**

# You and your friends are the future generation of oil people!

## Glossary

CASING PIPE - steel pipe used in oil wells to seal off fluids from the bore hole and to prevent the walls of the hole from sloughing off or caving.

CRUDE – unrefined petroleum.

DERRICK - the steel tower component of a drilling rig which supports the crown block, traveling block, and hoisting lines. Derricks are stationary structures normally requiring dismantling and disassembly when moved from location to location.

DISPOSAL WELL – many wells produce salt water along with oil. A disposal well is used to return produced water back into subsurface formations deep enough as to not harm shallower water sands.

DISCOVERY WELL – the first oil well drilled in a new field.

DRILL PIPE - A length of tube, usually steel, to which special threaded connections called tool joints are attached.

ENGINEER – designs the surface and subsurface activities related to the production of petroleum.

FOSSIL FUELS – oil, natural gas, coal formed naturally as part of the earth's processes, so-called because they are derived from the remains of ancient plant and animal life.
GEOLOGIST – a scientist who studies the structure, origin, history and development of the earth as revealed through the study of rocks, formations and fossils.
GEOPHYSICIST / GEOPHYSICS - the mapping of subterranean soil characteristics using non-invasive techniques such as sound.
GUSHER – an oil well which comes in at great pressure causing oil to jet out of the hole like a guyser.
PETROLEUM – crude oil or natural gas.
"PUMPJACK" PUMPING UNIT – mechanically lifts liquid out of the hole.
REFINERY – an industrial plant used to purify crude oil
RESERVES – discovered fossil fuels that are in the ground waiting to be brought to the earth's surface.
RESERVOIR – subsurface, porous, permeable rock surface generally containing three fluids; gas, oil and water.
SCOUT TICKETS - brief report about a well from the time it is permitted through drilling and completion. A scout ticket typically includes the location, total depth, logs run, production status and formation tops.
SEISMIC LINES - collection of data involves sending shock waves into the ground and measuring how long it takes for the subsurface rocks to reflect these waves back to the surface.
SPUD – to begin drilling; from the early days of the oil field using a spudder, also known as a cable tool rig.
TRAPPING MECHANISM - some sort of block to prevent the oil and gas from leaking away. Traps generally exist in predictable places - such as at the tops of anticlines, next to faults, in the updip pinchouts of sandstone beds, or beneath unconformities.
TANK BATTERY - an area where storage tanks are installed to receive produced fluids.

# For More Information

Ask Mom or Dad first before logging on.
For interactive crosswords, activity pages, helpful teacher guides, and a video book trailer go to:

## www.oilpeople.net

# Acknowledgements

The author wishes to thank the following for their knowledge and expertise:
Chris Bright, BS, Geologist, President, Sunlight Exploration, Inc.
Chance McMahan, BS, Geologist, Vintage Production California.
Greg Wilson, BS, MS, Geologist, President/CEO, Cogent Exploration, Ltd.

# Selected Bibliography

Abels, Jules; *The Rockefeller Billions*, The Macmillan Company: New York, 1965.

Conn, Frances G. & Rosenberg, Shirley Sirota; *The First Oil Rush*, Meredith Press: New York, 1967.

Giddens, Paul H.; *The Birth of the Oil Industry*, The Macmillan Company: New York, 1938.

Olien, Diana Davids & Roger M.; *Oil in Texas,* Austin University of Texas Press, 2002.

Parker, Charles; *The Oilmen*, Rinehart & Company, Inc.: New York, 1952.

Rintoul, Charles; *Drilling Through Time 75 Years with California's Division of Oil and Gas*, California Department of Conservation, 1990.

Tarbell, Ida M.; *The History of the Standard Oil Company,* W.W. Norton & Company Inc., 1969.

## Answer from Page 18:

Oil seeps naturally to the surface in many states, the most famous place being the La Brea Tar Pits in California. Oil can bubble out in plain sight or hide up to 16,000 feet below the earth's surface. It may be thick or thin, green or blackest black in color, smell sweet or sour, and it's all crude oil that can be refined into the products we use every day.

## About the Author

Natalie Bright is a freelance writer, creative writing instructor, and vice president for a small, independent oil and gas company. Married to a petroleum geologist for 25 years, they have two sons and live in the Texas Panhandle. www.nataliebright.com